Fire Boats

by Marcia S. Freeman

Consulting Editor:
Gail Saunders-Smith, Ph.D.

Consultant:
Mark Edelbrock
Fire Boat Pilot
Seattle Fire Department

Pebble Books

an imprint of Capstone Press
Mankato, Minnesota

Pebble Books are published by Capstone Press
818 North Willow Street, Mankato, Minnesota 56001
http://www.capstone-press.com

Library of Congress Cataloging-in-Publication Data
Freeman, Marcia S. (Marcia Sheehan), 1937–
    Fire boats/by Marcia S. Freeman.
    p. cm.—(Community vehicles)
    Includes bibliographical references and index.
    Summary: Describes fire boats, the equipment they carry, and the work they do.
    ISBN 0-7368-0101-4
    1. Fireboats—Juvenile literature. [1. Fireboats.] I. Title. II. Series.
TH9391.F74 1999
623.8'65—dc21   98-4250
                                                                            CIP
                                                                            AC

Note to Parents and Teachers

This series supports national social studies standards related to
authority and government. This book describes and illustrates fire
boats and the equipment they carry. The photographs support early
readers in understanding the text. The sentence structures offer
subtle challenges. This book introduces early readers to vocabulary
used in this subject area. The vocabulary is defined in the Words to
Know section. Early readers may need assistance in reading some
words and in using the Table of Contents, Words to Know, Read
More, Internet Sites, and Index/Word List sections of the book.

Table of Contents

Fire boats work on the water. Fire fighters use fire boats to fight fires on boats and docks.

6

Fire boats have horns and flashing lights. Horns and lights tell people that fire boats are coming.

Fire boats have pilot
houses. Fire fighters use
controls in pilot houses
to run fire boats. Fire boat
pilots steer fire boats.

Fire boats have monitors. Fire fighters use monitors to spray water on fires. Pumps draw water from under the boat.

Fire boats have hoses. Fire fighters use hoses to spray water or foam on fires. Foam puts out fires that water cannot put out.

search lights

FIRE
RESCUE

CHIEF SEATTLE

Fire boats have bright search lights. They light up the scene of a fire. Search lights help fire fighters find people in the water.

Some fire boats have dinghies. Fire fighters use these small boats to save people. Dinghies can go into places that large fire boats cannot reach.

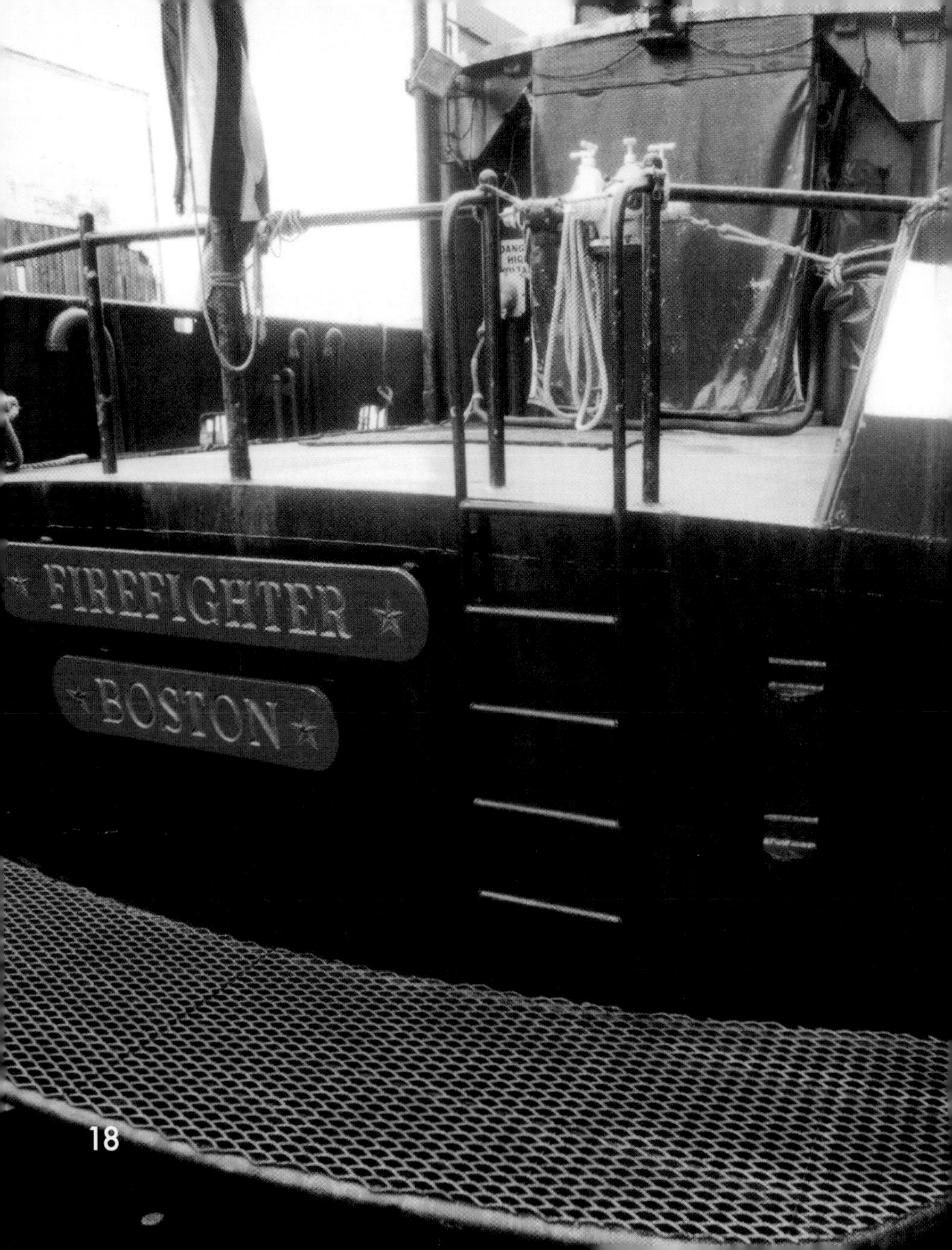
FIREFIGHTER
BOSTON

Fire boats have ladders.
Divers can go down the
ladders to the water.
Divers save people who
are in the water.

Fire boats dock at fire stations after fires. Fire fighters keep fire boats ready to fight fires.

Words to Know

controls—levers and switches used to make something work

dinghy—a small boat with a motor

foam—many small bubbles; fire fighting foam is made from a special liquid.

horn—a machine that makes a loud sound

hose—a long, bendable tube that carries water or foam from one place to another

monitor—a pipe through which water or foam is sprayed; monitors on fire boats can spray a column of water 295 feet (90 meters).

pilot house—the room on a fire boat that has the boat's controls

pump—a machine that forces liquid from one place to another; fire boat pumps force water or foam to monitors.

Read More

Gibbons, Gail. *Emergency!* New York: Holiday House, 1994.

Ready, Dee. *Fire Fighters.* Community Helpers. Mankato, Minn.: Bridgestone Books, 1997.

Somerville, Louisa. *Rescue Vehicles.* Look Inside Cross-Sections. New York: Dorling Kindersley, 1995.

Internet Sites

Fire Safe Kids Page
http://www.state.il.us/kids/fire

USFA Kids Homepage
http://www.usfa.fema.gov/kids/index.htm

Index/Word List

boat, 5, 11, 17
controls, 9
dinghies, 17
divers, 19
docks, 5
fire, 5, 11, 13, 15, 21
fire fighters, 5, 9, 11, 13, 15,
    17, 21
fire stations, 21
flashing lights, 7
foam, 13

horns, 7
hoses, 13
ladders, 19
monitors, 11
pilot houses, 9
pilots, 9
places, 17
pumps, 11
scene, 15
search lights, 15
water, 5, 11, 13, 15, 19

Word Count: 188
Early-Intervention Level: 10

Editorial Credits
Colleen Sexton, editor; Clay Schotzko/Icon Productions, cover designer;
    Sheri Gosewisch, photo researcher

Photo Credits
Dan Polin, 16
International Stock/Ron Frehm, 4
Mark C. Ide, 18
Mark Turner, 6, 8, 14
Michele Burgess, cover
Phillip Roullard, 12
PhotoBank, Inc./Mary Ann Hemphill, 10
Valan Photos/Arthur Burchell, 1; Ken Cole, 20